Mining Near Earth Objects

A Realization of Tomorrow

Robert Krueger

Mining in outer space is an industrialization that will propel Earth's global economy into a new paradigm. With untapped resources able to sate the global production demands for millions of years, this new industry will provide an economic boon of unprecedented measure that will usher humanity into a brand new age.

Earth's supplies of rare and precious resources are quickly depleting. Many of our heavy metals will be exhausted in 50 to 60 years; the rarest and highest in demand will be exhausted within the next 20 years. The most abundant supplies of these rare resources currently exist outside of our planet in space. Mining in outer space is a future necessity—it will provide an economic boon that will promote industry, drive innovation and increase space accessibility.

While our supply of precious resources here on Earth are depleting, our demand for them continues to rise. One of the resources highest in demand and lowest in supply is Helium-3(^{3}He), which is used in nuclear fusion reactions. As new terrestrial sources are sought, materials are obtained at increasing economic and environmental cost. Society pays for this depletion of resources in the form of higher prices for manufactured goods, would-be technologies that are not developed for lack of raw materials, global and regional conflicts spurred by competition for remaining resources, and environmental damage caused by development of poorer and more problematic deposits.

Finding new and easily exploitable sources is the first step towards establishing a profitable space-based economy. According to McKay & McKay (1992), utilization of asteroid resources may provide a strong solution to the problem, as they hold the potential for becoming primary sources of some precious metals and other materials. Precious metals and semiconducting elements in iron meteorites, which commonly form the metallic cores of asteroids, are found in very large concentrations compared to Earth sources. For instance Ostro (1985) in his radar study states that 16 Psyche, a

200km diameter M-type asteroid, appears to be a fairly pure composition of iron-nickel. This could supply the world with a surplus that meets production requirements for several million years.

Asteroid: 16 Psyche - Image credit: NASA/JPL-Caltech

Studies throughout the years have indicated that ^{3}He will be a very important resource in the future as fusion reactors become more prominent. Taylor & Kulcinski (1999) wrote about the future state of earth's natural energy resources, the potentials of ^{3}He and applications it will have in the future of humanity. "There is sufficient ^{3}He in the upper 3 meters of only the maria of the Moon to supply the entire energy needs of the Earth for over a thousand years. The potential for greatly enhanced supplies of ^{3}He at the lunar poles may make the utilization of this energy-generation process even more attractive" (p. 33). The abundance of ^{3}He on the lunar surface has been theorized and then mapped. Johnson et al. (1999) correlated locations of estimated high ^{3}He concentration to solar winds acting upon the landscape with data collected from Apollo landing sites. After analyzing this information, theory for all of the variables necessary for high concentration ^{3}He deposit development has been proposed. Fa and Jin (2007) estimate that the lunar inventory of ^{3}He in just the regolith layer is 3.72×10^8 kg on the nearside and 2.78×10^8 kg on the farside. Whereas Earth's global inventory of ^{3}He is estimated at less than 1×10^3 kg in entirety. Since ^{3}He is considered one of the most valuable resources for the future of mankind, seeking it out and exploiting it should be priority. Two space-based commercial markets other than mining are projected to come online within the next few decades, namely solar power stations, and space tourism. The concept of space based solar power (SBSP) received renewed consideration—the Japanese have considered an equatorial orbit SBSP pilot plant, orbiting at 1100 km altitude, of mass 200 tonnes (Nagatomo, 1996). Japan's National Space Development Agency, hoping to launch an experimental version of an SBSP within the next decade, has asked several private companies to submit design proposals. The feasibility of space tourism is also being promoted. These power plants would produce electricity through photovoltaic cells and beam the abundance to earth ground stations through any of various proposed mediums such as laser, electromagnetic, or microwave transmission.

Collins' (1994) study found the following:

> Market research in the United States, Japan, Canada and Germany has shown that as many as 80% of people younger than 40 would be interested in commercial space travel. A majority would be willing to pay up to three months' salary for the privilege. Ten percent would pay a year's salary. It is estimated that at a launch cost of

$200/kg the space tourism industry will grow rapidly to several billion dollars per year. (p. 5)

Hotels in orbit will be needed to cater for large accommodations within the next decade; estimated market demands are as large as 10,000 people. The Japanese Shimizu Corp., an engineering and construction firm, has developed a plan for an orbiting hotel to fulfill this estimated future need. They have pledged to have their 6000 tonne, 64-room hotel in orbit by 2020; they further plan to establish a lunar unit shortly after. As a result of these and other activities, we can expect a future market for materials in LEO—particularly construction supplies, station-keeping and transportation propellants, and shielding against cosmic radiation. "The size and rate of this future in-orbit market for materials could exceed 1000 tonnes per year by 2010, growing exponentially to tens of thousands of tonnes per year if any large scale activity develops rapidly" (Sonter, 1997).

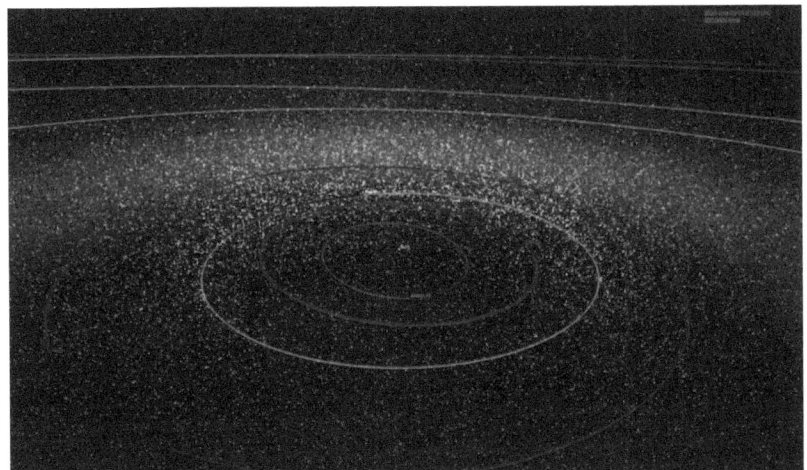

This image depicts a mapping of the positions of known near-Earth objects (NEOs) at points in time over the past 20 years. Asteroid search teams supported by NASA's NEO Observations Program have found over 95 percent of near-Earth asteroids currently known. There are now over 18,000 known NEOs and the discovery rate averages about 40 per week. Image credit: NASA/JPL-Caltech

Much of the material needs of these markets could potentially be offset by Asteroid mining. Asteroid mining is a concept that involves the extraction of useful materials from asteroids. Due to their accessibility, near-Earth asteroids (also known as NEAs) are a particularly accessible subset of the asteroids that provide potentially attractive targets for resources to support space industrialization. Many materials could be extracted and processed from NEAs which are useful for propulsion, construction, life support, agriculture, metallurgy, semiconductors, and precious and strategic metals. Precious rare resources can be found in very large concentrations compared to Earth sources. From NEA sources, we can expect high extraction rates "up to 187 parts per million (ppm) of precious metals... more than 1000 ppm of other metals, semiconductors, and nonmetals may one day be extracted and imported by Earth from asteroids." (Ross, 2001). By fulfilling current and future resource demands, the industrialization of space is postured to be very lucrative.

    The task of establishing profitable space industrialization will be a complex realization. In order to ensure minimized risk and maximized profit, much designing, planning and engineering will take place beforehand. Erickson (2006) outlines an example mission to a near Earth asteroid by following these concepts with an emphasis on flexibility, cost-effectiveness and fail-safes to maximize success and ensure a positive return-on-investment. His plan is to bring all the equipment and personnel required to land on an asteroid, mine the raw materials, refine the materials, and then return the refined materials to earth. This first method is the most complex.

The second potential approach is to mine a NEA and return raw material for processing. Gertsch and Gertsch (2000) proposed a project scaled to the equivalent of the Chunnel. They estimate that an endeavor of this scope would cost at least $5 billion and require about 12 years to complete. This study assumes that the asteroid mined would consist of 150 parts per million of Platinum-group metals, a concentration thought to occur in about one in 10 platinum-bearing asteroids. Finding a suitable asteroid and mounting a mission could consume up to four years of the project. Upon arrival, miners would need to sift through about 500 million metric tons of material in order to extract enough unrefined platinum—68 thousand metric tons at an assumed price of about $13 per gram—to break even on the project. Though, a full return on investment would not attract billions in risk capital especially considering the 12-year timeframe and the moderate likelihood of failure. A third method has recently been proposed which consists of bringing smaller NEAs in entirety to Low Earth Orbit (LOE) which will there be mined and refined. Oleson & Brophy (2012) examine this approach with a proof-of-concept example: a mission to retrieve 2008 HU4. The concept is to launch a high-power solar electric propulsion system (SEP) into LOE. Once all necessary components are in orbit, they would be assembled into a travel configuration where it would then begin journey towards a rendezvous with asteroid HU4. Upon arrival to HU4, the SEP would begin to orbit and then decelerate around HU4, match speed rotation and precession to the object, enter a landing configuration and begin a controlled descent to the far surface. Once on the surface, the SEP will anchor, reenter travel configuration and begin altering the asteroid's course for a Trans-lunar injection from which it will further be altered to a LOE. Between launch, attaching a high-power solar electric propulsion system to HU4 and bringing this asteroid into LOE; the mission length would be approximately 10 years. "The enormous reduction in initial mass in low-Earth orbit enabled by SEP would make an asteroid retrieval mission affordable for the first time in history. This has the potential to jump-start the in situ resource utilization industry" (Oleson & Brophy, 2012).

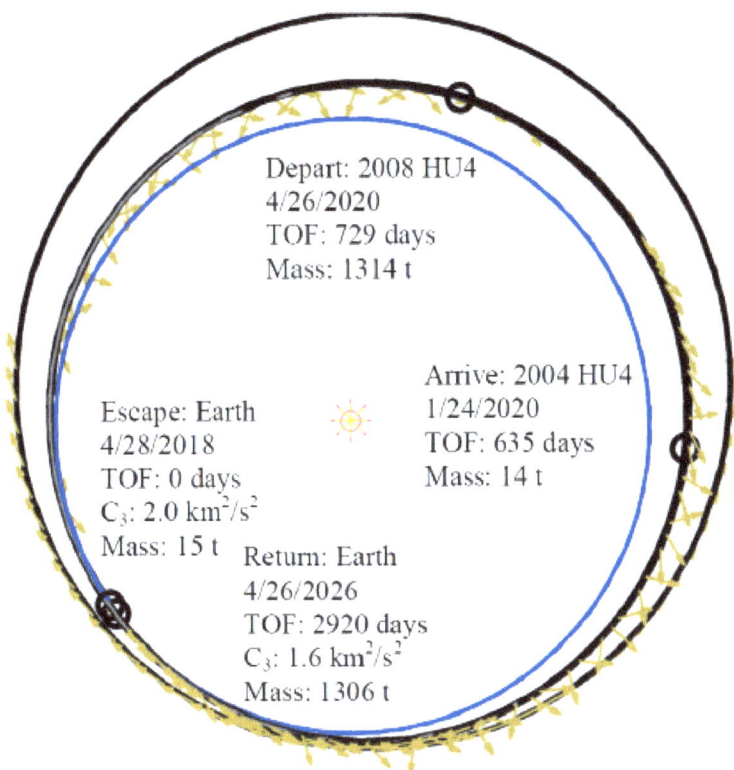

Example mission returning 2008 HU4, a small (~7 m), 1300 t of NEA with a radar opportunity in 2016.

The level of planning and preparation necessary for initiating a new high tech and high profit industry facilitates a strong drive for innovation. This drive is even more pronounced when analyzing the unconventional approaches outlined to reduce launch costs.

Accessing space is the greatest hurdle profitable space mining faces. Utilizing current techniques, the margins on a positive return on investment are very slim. In order to make space mining a more secure investment to ensure it gets the funding it needs, the cost of getting personnel and equipment to LOE will need to fall significantly.

The current cost of putting a payload into LOE is exorbitant. At about $10,000 per kg, there isn't much room for trial and error. Just lifting a payload equal to the mass of Hubble-telescope costs $115,740,000. A single tunnel boring machine, used to tunnel mines all around the world, would cost around $20.5 billion to lift into LOE. According to Powell, Maise, and Rather (2010), "Human access to the ISS costs $20 million for a single passenger" (p.1).

Powell, Maise, and Rather (2010) explore the possibilities of maglev space access (StarTram). By applying the energy requirements more directly upon the weight of the payload itself, energy costs can effectively be reduced to as low as 53 cents per kilogram. Magnetic Levitation (Maglev) vehicles should be able to make this a reality. Maglevs are magnetically levitated above a guideway without mechanical contact or friction. Current Maglev vehicles operating in Japan are capable of reaching speeds of 360 mph; the only limiter keeping them from going faster has been air drag.

In order to apply the energy required to reach LOE upon the payload, there needs to be a reduction of mass on the launch vehicle. StarTram does this through locating their expensive and complex equipment on the ground. This is completely contrary to the current method of launch where all expensive and complex equipment is located aboard the rocket. By doing this, the weight of the escape vehicle needs is reduced significantly.

Citing more than a 10,000 times reduction in payload cost, these potential launch systems would make space much more accessible. Effectively, a ticket to outer space would drop from $20 million to less than $2,000. Cargo would be able to get into LOE for less than one dollar per kilogram, dropping the total cost of lifting a single tunnel boring machine to only $2 million.

By switching over to a Maglev launch system, we'll be able to more efficiently deliver payloads to orbit; reducing energy costs to as low as 53 cents per kilogram, about 10,000 times smaller than present launch costs. Maglevs are magnetically levitated above a guide-way without mechanical contact or friction. Current Maglev vehicles operating in Japan are capable of reaching speeds of 360 mph; the only limiter keeping them from going faster has been air drag. By using a low pressure tunnel, Maglev speed will be practically limitless. Once escape velocity is achieved the Maglev will enter the atmosphere at high altitude, lower air density, towards orbit. The only assistance required would be a short delta v burn by a small rocket attached to the spacecraft.

All current launch vehicles have their expensive and complex launch equipment located on the vehicle. Conversely, StarTram will be locating all of its expensive and complex equipment on the ground. The ground equipment will be reusable, capable of launching several craft. Current rockets launch but a few hundred tons of payload every year and are then disposed of. A StarTram site will be capable of putting thousands of tons of payload into orbit per year and at a significantly reduced cost. This will vastly increase our interactions with space allowing greater access to exploration and commercial pursuits.

For example, one StarTram launch site would be able to place 100 gigawatts of space solar power satellites every year at a launch cost of only 200 dollars per PW. This is a small fraction of what current earth power plants cost to operate. In only 10 years, space solar power could provide one third of the world's energy needs. This would more than make up for any costs associated with StarTram's investment.

A second accessibility concern is access to more distant space locations. With current propulsion solutions the travel time for missions to closer celestial bodies, such as Mars, takes several months. New propulsion technologies will need to be developed and tested to increase the feasibility of retrieving more distant resources and reducing the mission time significantly. Significant testing needs to be conducted in this field, the project best postured for this task was cancelled. "Project Prometheus was established to develop technologies that enable a new class of deep space missions that cannot be achieved with chemical propulsion. Project Prometheus' goal is to develop the first reactor powered spacecraft and demonstrate it can be safely operated for long periods of time on deep space missions" Whiffen (2003). In order to make more distant locations accessible and to reduce mission times, Project Prometheus needs to be funded once more so our research into advanced space propulsion can continue to advance.

Space accessibility also extends to the laws in place which govern responsibility and liability in outer space. Lee (2012) notices an absence of appropriate legal framework. these legal and policy issues regarding the concept of commercial utilization of outer space resources.. His research includes analysis on economic necessity, technical feasibility, financial requirements, economic risk and legal risk. After going into depth on minimizing the legal risks involved in order to finance space mining projects, he evaluated the current state of international space law, namely: state responsibility, international liability, extraction and mining rights, and celestial body sovereignty. The two existing international legal frameworks regarding space law fall severely short of this task. The Outer Space Treaty, signed in 1967, lays out a ban on weapons of mass destruction in outer space, also calling for a demilitarized space free of sovereign claims. This treaty is signed and observed by 102 countries. The Moon Agreement, signed in 1979, was an attempted treaty to assign jurisdiction of celestial bodies to the international community. Further clauses included: bans of military use on celestial bodies, ban of unapproved celestial body exploration, Secretary-General notification of all celestial activities and discoveries, sample sharing with international community, contamination ban, ban of celestial body sovereignty, ban of extraterrestrial property ownership outside of international governmental, and all resource extraction must be made by international organizations. This agreement has only been ratified by 15 states, none of which possess a space program (Dembling & Arons, 1967).

With the realization of establishing space based mining, we will be able to more easily access outer space through new technologies available through StarTram and Prometheus, drive innovation with plans and designs promoting problem solving on a massive scale to increase ROI and minimize risk, and promote industry through injecting a supply of materials to alleviate costs of future endeavors and to sate a global market that is running on fumes. This will greatly stimulate the Earth's economy and replenish our nearly depleted and depleted resources. The space option is humanity's most optimistic approach to the future, and space mining is just the beginning.

433 Eros is a stony asteroid in a near-Earth orbit
Image Credit: NASA/NEAR Project (JHU/APL)

This representation of Ceres' Occator Crater in false colors shows differences in the dwarf planet's surface composition.
Image Credit: NASA/JPL Caltech/UCLA/MPS/DLR/IDA

Artist's impression of NASA's New Horizons spacecraft encountering 2014 MU69, a Kuiper Belt object that orbits the Sun 1 billion miles (1.6 billion kilometers) beyond Pluto, on Jan. 1, 2019.
Image Credit: NASA/JHUAPL/SwRI

This artist's concept depicts the spacecraft of NASA's Psyche mission near the mission's target, the metal asteroid Psyche.
Image Credit: NASA/JPL-Caltech/Arizona State Univ./Space Systems Loral/Peter Rubin

In this artist's concept, a narrow asteroid belt filled with rocks and dusty debris orbits the sun.
Image credit: NASA/JPL-Caltech

Brahin is a meteorite pallasite found in 1807. This is the second meteorite ever found in Russia. Sometimes it is also called Bragin or Bragim. It is quite common among collectors due the affordable price of small partial slices.
Image Credit: Steve Jurvetson

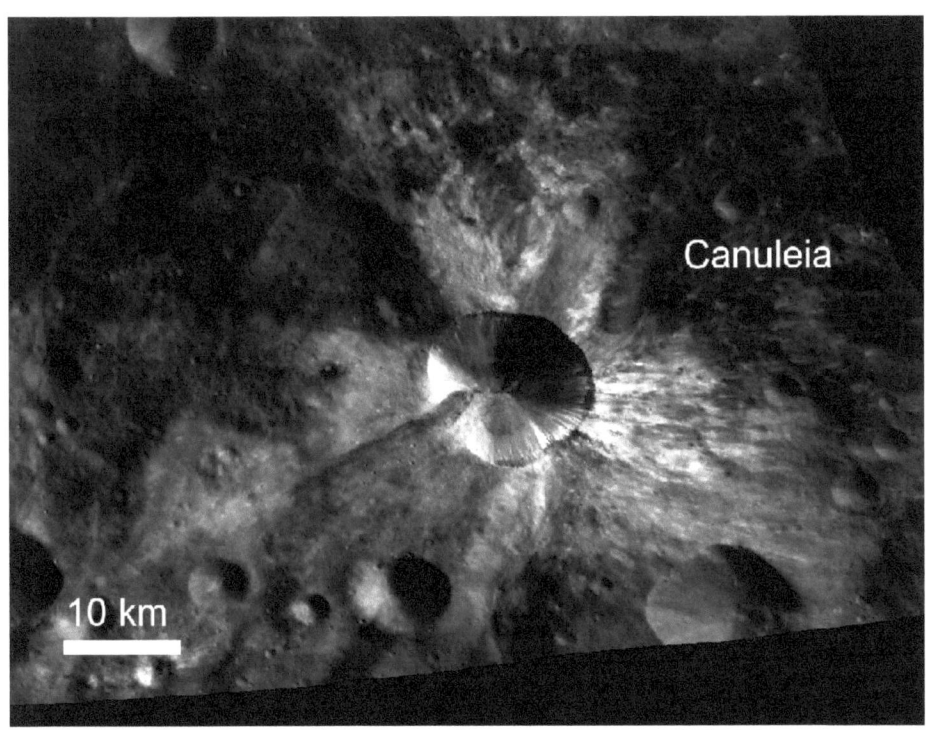

In this image from NASA's Dawn spacecraft, bright material extends out from the crater Canuleia on Vesta. The bright material appears to have been thrown out of the crater during the impact that created it. Image credit: NASA/JPL-Caltech/UCLA/MPS/DLR/IDA/UMD

Works Cited

Brenan J. M. & McDonough W.F (2009) Core formation and metal–silicate fractionation of osmium and iridium from gold. *Nature Geoscience*.

    http://www.geology.utoronto.ca/Members/brenan/Brenan%20and%20McDonough,%202009https://www.jpl.nasa.gov/wise/newsfeatures.cfm?release=2018-259.pdf

Dembling, P. G., & Arons, D. M. (1967). The Evolution of the Outer Space Treaty.

    http://digitalcommons.unl.edu/cgi/viewcontent.cgi?article=1002&context=spacelawdocs

Erickson, K. R. (2006). Optimal architecture for an asteroid mining mission: equipment details and

    integration. *Space*.

    http://arc.aiaa.org/doi/abs/10.2514/6.2006-7504

Fa, W., & Jin, Y. Q. (2007). Quantitative estimation of helium-3 spatial distribution in the lunar regolith

    layer. *Icarus*, *190*(1), 15-23.

    http://www.sciencedirect.com/science/article/pii/S0019103507001285

Gertsch, L. S., & Gertsch, R. E. (2000, January). Mine planning for asteroid orebodies. In *Space*

    Resources Roundtable II (Vol. 1, p. 19).

    http://adsabs.harvard.edu/abs/2000srrt.conf...19G

Johnson, J. R., Swindle, T. D., & Lucey, P. G. (1999). Estimated solar wind-implanted helium-3

    distribution on the Moon. *Geophysical research letters*,*26*(3), 385-388.

    http://onlinelibrary.wiley.com/doi/10.1029/1998GL900305/abstract

Lee, R. (2012). *Law and Regulation of Commercial Mining of Minerals in Outer Space*. Springer Verlag.

    http://books.google.com/books?hl=en&lr=&id=l1Zu4K7_VC0C&oi=fnd&pg=PR7&dq=asteroid+mining+innovation&ots=FWMB_GqAb6&sig=JWSBiS-

RwkRh3atcRB6MMh3Bzms#v=onepage&q=asteroid%20mining%20innovation&f=false

McKay, M. F., McKay, D. S., & Duke, M. B. (1992). *Space resources. Volume 3: Materials* (No. N-93-16875; NASA-SP--509-VOL-3; S--689-VOL-3; NAS--1.21: 509-VOL-3; LC--92-4468). National Aeronautics and Space Administration, Houston, TX (United States). Lyndon B. Johnson Space Center.
    http://catalog.lib.byu.edu/uhtbin/isbn-search/0160380626

Oleson, S., & Brophy, J. R. (2012). Spacecraft conceptual design for returning entire near-earth asteroids.
    http://trs-new.jpl.nasa.gov/dspace/bitstream/2014/42753/1/12-3200_A1b.pdf

Ostro S. J (1985) Radar observations of asteroids and comets. *Astronomical Society of the Pacific.*
    http://articles.adsabs.harvard.edu/cgi-bin/nph-iarticle_query?db_key=AST&bibcode=1985PASP...97..8770&letter=.&classic=YES&defaultprint=YES&whole_paper=YES&page=877&epage=877&send=Send+PDF&filetype=.pdf

Powell, J., Maise, & G., Rather, J. (2010, January). Maglev Launch: Ultra-low Cost, Ultra-high Volume Accessible to Space for Cargo and Humans. In *AIP Conference Proceedings* (Vol. 1208, No. 1, p. 121).
    http://www.physics.utu.fi/projects/kurssit/UFYS3032/StarTram2010.pdf

Ross, S. D. (2001). Near-Earth asteroid mining. *Space.*
    http://www.isdc2007.org/settlement/asteroids/NearEarthAsteroidMining(Ross2001).pdf

Sonter, M. J. (1997). The technical and economic feasibility of mining the near-earth asteroids. *Acta Astronautica, 41*(4), 637-647.
    http://ro.uow.edu.au/cgi/viewcontent.cgi?article=3862&context=theses

Taylor, L. A., & Kulcinski, G. L. (1999). Helium-3 on the Moon for fusion energy: The Persian Gulf of the
    21st century. *Solar System Research*, *33*, 338.
        http://adsabs.harvard.edu/abs/1999AVest..33..338T

Whiffen, G. J. (2003). An investigation of a Jupiter Galilean moon orbiter trajectory.
    http://trs-new.jpl.nasa.gov/dspace/bitstream/2014/38513/1/03-2062.pdf

Jurvetson, Steve. The Rich History Of the Brahin Pallasite
    https://www.flickr.com/photos/jurvetson/6098963730

Jet Propulsion Laboratory, California Institute of Technology. Asteroids, Comets, Planets: Cut From Same Cloth?; Cosmic Detective Work: Why We Care About Space Rocks;
    https://www.jpl.nasa.gov/news/news.php?feature=1199
    https://www.jpl.nasa.gov/wise/newsfeatures.cfm?release=2018-259

www.ingramcontent.com/pod-product-compliance
Lightning Source LLC
Chambersburg PA
CBHW040341220526
45473CB00009B/2761